BY KEN CAMERON

## Table of Contents

CHAPTER 1

# So Many Different Plants

Have you ever noticed how many different kinds of plants there are in the world? Some grow to be tall trees. Others creep along the ground. Some are prickly. Others are smooth. Some live only a short time. Others live hundreds of years. Some are poisonous, and others are good to eat.

Why are there so many different kinds? The answer can be found inside the stems, roots, leaves, and flowers of each plant. If you cut open a plant stem or leaf and look at it under the lens of a microscope, you would see that the stem and the leaf are made up of hundreds of tiny, square-shaped compartments. These compartments are called **cells**.

cushion plants

← tropical shrub cactus ↓

## The nucleus is like the brain or command center of the cell.

Within every one of these microscopic cells is an even smaller ball-shaped structure called the **nucleus** (NOO-clee-us). The nucleus is like the brain or command center of the cell. It contains all the instructions that tell the plant how it should grow, what color it should be, how tall it should get, and what it should taste like. The instructions inside the nucleus also decide whether the plant will be poisonous or good to eat.

Onion root tip cell

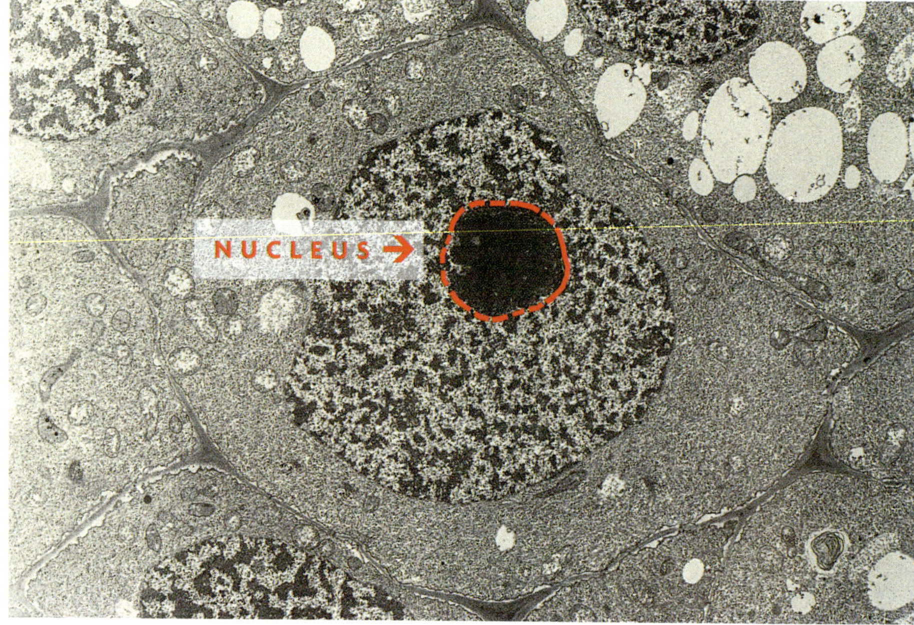

When a seed first begins to grow, it "knows" exactly what kind of plant it will become. It has a special set of instructions within the nucleus of each one of its cells. Plants that are the same have the same set of instructions. A more scientific name for this set of instructions is the plant's **genes** (pronounced like "jeans" that you wear). The study of plant genes is called plant **genetics** (jen-ET-iks). A scientist who studies plant genetics is called a plant geneticist (jen-ET-iss-ist).

**This seedling, or baby plant, contains all the genes to become an adult bean plant.**

CHAPTER 2

# The Story of the First Farmers

Thousands of years ago people did not have farms. Instead, they got food by hunting animals and gathering plants that grew wild.

One plant they gathered was a type of wild grass. They collected its small seeds and ground them into flour to make a kind of bread. That wild grass was the **ancestor** of what we now call wheat.

This scene shows Egyptian farmers planting seeds.

At some point, scientists believe, someone had the idea to save some of the seeds from the wild grass plants and put them in the ground. This person was the world's first farmer. From those seeds, more plants grew. Now people didn't have to go out looking for wild plants in order to get the seeds they needed to make bread. They could make the plants grow where they wanted them. This was the beginning of **agriculture**.

**QUESTION:**

**How would your life change if you had to search for food each day instead of opening a refrigerator door?**

**After thousands of years of selecting, or choosing, the biggest seeds, farmers ended up with what we know today as wheat.**

Scientists also think that at some point, an early farmer noticed that some of the wild grass plants made larger seeds than others. Larger seeds were better than smaller seeds, because not as many seeds were needed to make the same amount of flour. The farmers picked out these larger seeds to plant. Then more plants with large seeds grew. The instructions for making plants with larger seeds was in the nucleus of each cell in the plant.

## HISTORY LESSON

**Farming began in an area in the Middle East near the Tigris and Euphrates Rivers. Today we refer to this area as "the Fertile Crescent." Ancient grains of wheat found in this region have been estimated to be more than 10,000 years old.**

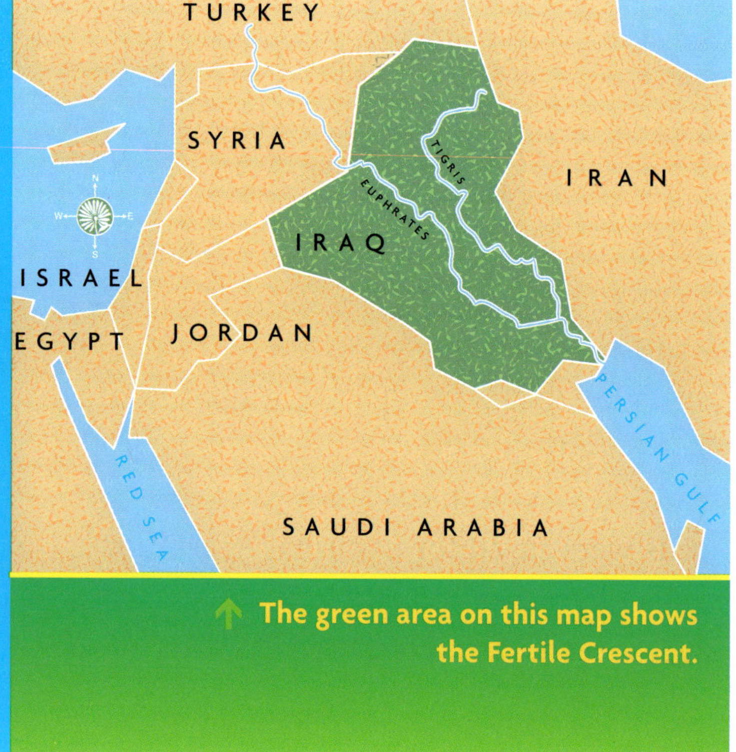


The green area on this map shows the Fertile Crescent.

**THINK IT OVER!**

**When we think of wheat, we usually think of bread.**

**Make a list of other foods with wheat in them.**

After planting more and more of these large seeds each year, the ancient farmers would sometimes find one or two plants in their fields that had seeds that were a little bit bigger than the others. They used most of the seeds for flour, but kept the largest seeds they could find each year. Those seeds were saved for planting. After thousands of years of selecting, or choosing, the biggest seeds, farmers ended up with what we know today as wheat. It came from nothing more than an ordinary grass.

**It took thousands of years of choosing the corn grass plants with the biggest seeds, or kernels, to make what we enjoy today as corn on the cob.**

The early farmers didn't realize it at the time, but each year they were choosing the grass plants with the best set of genes—genes that carried instructions for making larger seeds. Although they didn't know it, they were using the science of plant genetics.

Many other crops grown by farmers today have been selected from wild plants in the same way. For example, have you ever found and eaten wild strawberries? They are much smaller than strawberries that are grown on farms. How do you think the larger strawberries that you buy in the store came from the tiny wild strawberries that you can pick in the woods?

wild strawberries

domestic strawberries

Would it surprise you to know that corn is also a type of grass? It was first grown in Central America thousands of years ago. Its seeds are called kernels. Very few kernels grew on wild corn grass. It took thousands of years of choosing the corn grass plants with the biggest seeds, or kernels, to make what we enjoy today as corn on the cob.

→

**Modern corn plants have larger kernels than wild corn grass.**

Here's another example of how plant genetics has changed the food we eat. Did you know that cabbage, broccoli, cauliflower, Brussels sprouts, and many other vegetables all came from the same wild plant? To create cabbage, farmers selected plants that made one superlarge bud in the middle. To create Brussels sprouts, they chose plants that made many smaller buds along the stem. Cauliflower is just a type of broccoli that doesn't become green. Its genes are just a little bit different.

↑ Brussels sprouts

↓ cauliflower

← broccoli

CHAPTER 3

# The Science of Plant Genetics

For a long time, people continued to select seeds from plants in nature that would grow the best food on their farms and the prettiest flowers in their gardens. They still didn't know that they were selecting plants with the best genes, but that is what they were doing.

They also learned how to create new kinds of plants by **crossing** one kind of plant with another. The flowering parts of the plants are important in this process.

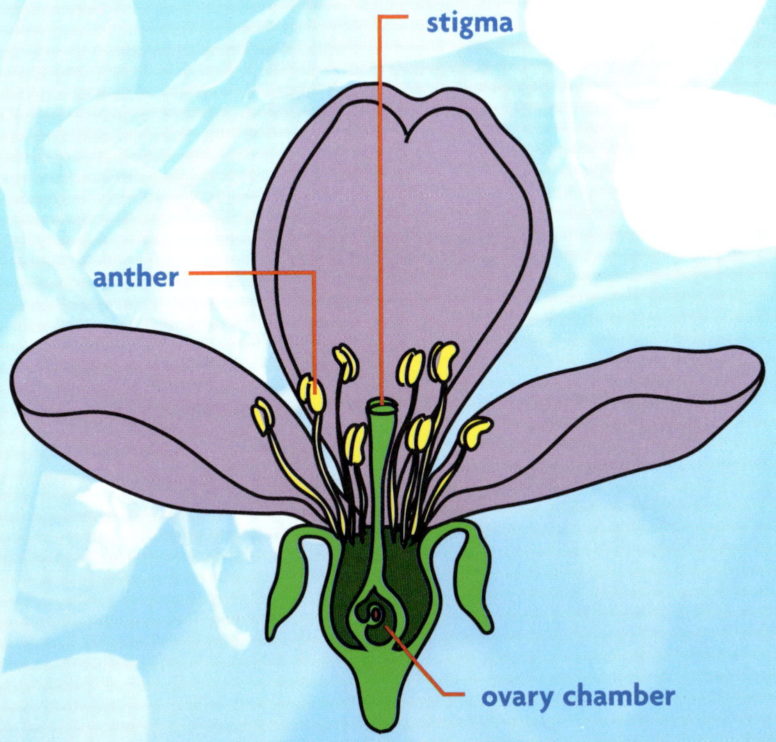

## A plant that is the result of crossing two other plants is called a hybrid.

To cross two kinds of plants, you take some of the dust-like, yellow **pollen** from the **anther** of one plant's flower and put it onto the sticky, green **stigma** of the flower of a different kind of plant. Weeks later, seeds should develop in the **ovary chamber** of the plant that received the pollen. When those seeds are planted, they will grow into a kind of plant that is a mixture of the original two plants. The new plant will have some genes of one plant and some genes of the other. A plant that is the result of crossing two other plants is called a **hybrid**. Two examples of hybrid fruits are the loganberry and the boysenberry.

The loganberry is a cross between a blackberry and a raspberry. →

The boysenberry is a cross between a blackberry and a loganberry. →

Have you ever eaten a tangelo? The tangelo is a cross between a grapefruit and a tangerine. Many people love the sweet taste of tangerines, but tangerines are not very big. Grapefruits are bigger, but they are not very sweet. By crossing a grapefuit with a tangerine, farmers have been able to create a larger fruit that has the sweetness of a tangerine.

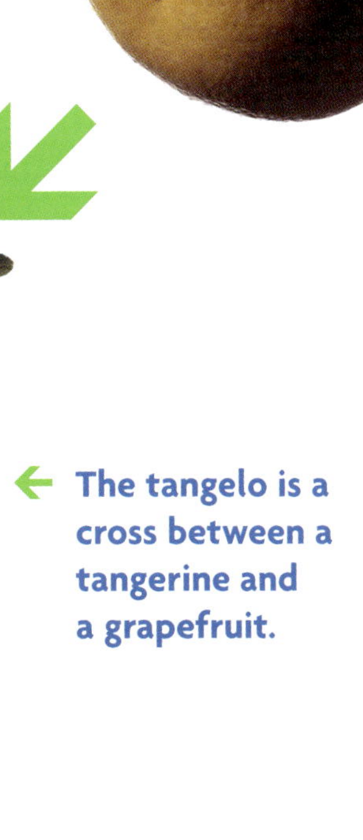

The tangelo is a cross between a tangerine and a grapefruit.

**COMMON NAME:** Tomato

**SCIENTIFIC NAME:** *Lycopersicon esculentum*

**COUNTRY OF ORIGIN:** Peru

**INTERESTING FACTS:** When the first Europeans saw tomatoes in 1544, they would not eat them. They believed the red tomato fruit was poisonous. Now tomatoes are eaten in almost every country of the world.

Some hybrids, such as the tangelo, are wonderful because they combine the best genes of two different kinds of plants. Other hybrids are not so wonderful.

For example, farmers once tried to cross a tomato and a potato plant. Potatoes grow underground. Tomatoes grow above ground.

Farmers wanted to create one plant that would grow potatoes underground and tomatoes above ground. Instead, the hybrid potatomato plant made roots like a tomato underground and stems like a potato above ground. It didn't produce potatoes or tomatoes. The hybrid had the worst genes of both plants blended together.

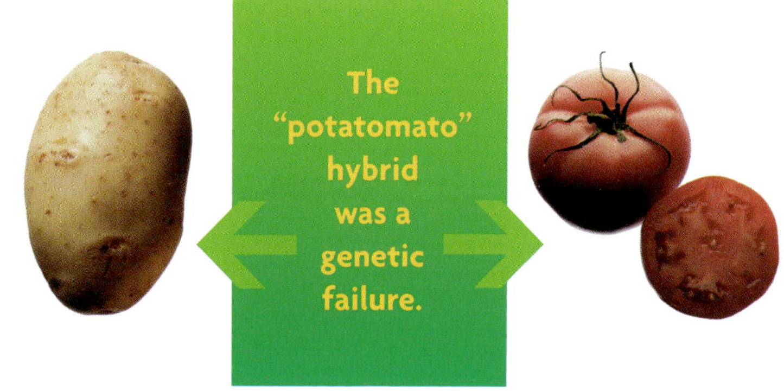

The first person to experiment with crossing plants was Gregor Mendel. Mendel was a monk who enjoyed gardening. He became curious about why some pea plants were tall and others were short, and he decided to experiment. Although he did not think of himself as a scientist, he certainly acted like one. He set up very careful experiments and he kept complete notes. He accurately recorded all his results.

Gregor Mendel was born in Austria in 1822.

pea plant

**Mendel found that if you crossed a tall pea plant with a short pea plant, all the new seeds would produce tall plants.**

Mendel began his experiments with four steps and then repeated them many times. First, he crossed short and tall pea plants. Second, he removed the seeds from the plants a few weeks later. Third, he planted the new seeds. Fourth, he wrote down the height of the new plants.

Mendel found that if you crossed a tall pea plant with a short pea plant, all the new seeds would produce tall plants. However, when he crossed these new tall plants with each other in a second experiment, some of the seeds from those tall plants produced short plants!

**IT'S A FACT!**

**Did you know that the peas we eat today are really the seeds of the garden pea plant?**

In order to understand why this happened, you need to know that there are two genes for each characteristic of a plant, such as its height or color. You also need to know that certain genes are stronger than others. For example, the gene for tallness in pea plants is stronger than the gene for shortness. Today we call the stronger genes **dominant** and the weaker ones **recessive**.

If a pea plant has two tall genes, of course it will be tall. If it has one tall gene and one short gene, it will still be tall, because the tall gene is stronger than the short gene. In order for a pea plant to be short, it must have two short genes. If you cross a tall pea plant with a short pea plant, all the plants you get will have one tall and one short gene. But the plants will all be tall.

In this diagram, the plant with two tall genes is called a "TT" plant; the plant with one tall and one short gene is called a "Ts" plant; and the plant with two short genes is called an "ss" plant. (A capital "T" is used to show that tall is dominant and a small "s" is used to show that short is recessive.)

But say you have two tall pea plants. Each one has one tall gene and one short gene. You cross the two plants, and you plant four of the seeds that you get. The chances are that you will end up with one plant that has two tall genes, two plants that have one tall and one short gene, and one plant that has two short genes. All the plants will be tall except for the one with the two short genes. That plant will be short.

Crossing two pea plants that have one tall gene and one short gene produces four possible genetic combinations.

Mendel carefully wrote down everything he did in his experiments with garden peas. Then, in 1865, he wrote a book called *Experiments in Plant Hybridization*. It was the first book about plant genetics.

**IT'S A FACT!**

**Few people read Mendel's book until after he died. It wasn't until then that he became famous. Now he is called the Father of Genetics.**

Gregor Mendel's garden in the former Czechoslovakia

CHAPTER 4

# The Importance of DNA

Mendel's experiments were an important first step in understanding plant genetics, but later scientists still had many questions. For example, what are genes made of, anyway?

We know now that genes are made of DNA. You may have heard of something called DNA, but you may not understand just what DNA is or why it is so important. In the 1940s, researchers discovered that DNA was a chemical substance. But it was not until 1953 that two scientists working in England, James Watson and Francis Crick, discovered the actual structure of DNA. The letters "DNA" stand for deoxyribonucleic (dee-ahks-ee-RYE-boh-noo-KLAY-ik) acid.

**James Watson (left) and Francis Crick (right) along with another scientist, Maurice Wilkins, received the Nobel Prize in Medicine in 1962.**

Watson and Crick discovered what DNA looks like and how it works. To imagine what DNA looks like, picture beads on two long strings that are joined like the sides of a ladder and twisted together. Imagine that the beads come in four colors, arranged in different patterns on the strings. Of course, DNA is not really made of beads. It is made of four different kinds of **molecules** called A, T, C, and G. The letters stand for four different chemicals.

**IT'S A FACT!**

**The names of the chemicals in DNA are**

- **adenine (ADD-en-een)**
- **thymine (THYE-mean)**
- **cytosine (SYE-toh-seen) and**
- **guanine (GWAH-neen).**

This is what a DNA strand looks like.

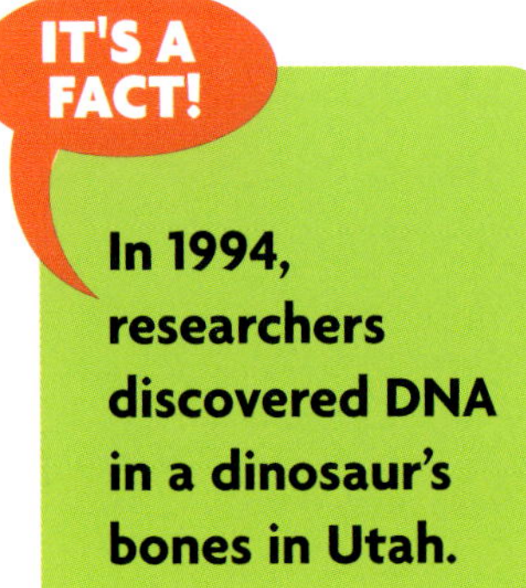


It is the sequence, or order, of these four molecules on the strings of DNA that makes each gene special. There is a special gene for every characteristic of every plant, from flower color, to leaf size, to height. Every gene is like a message in secret code that tells the plant how to grow in one special way.

For example, one section of a plant's DNA string contains the gene for color. If the molecules in that section are arranged A-G-A-C-C-C, the plant might have red flowers. If they are arranged T-G-A-C-T-T, the plant might have white flowers.

← DNA includes four molecules (A, T, C, and G) represented here as yellow, green, red, and orange.

# The Future of Plant Genetics

We've come a long way since Gregor Mendel did his first experiments with garden peas. Just recently, a team of scientists from all over the world made the first complete list of every gene in the nucleus of a plant called the mouse-eared cress. But we still don't know everything about how genes and DNA work.

For example, a flowering plant's genes may tell it to make a flower, but without a sunny environment, the plant won't do it. Scientists don't yet fully understand the reason for this. In other words, although a plant may have the genes that tell it how to grow, the genes won't be "turned on" unless the plant is in the right surroundings.

**PLANT PROFILE**

**COMMON NAME:** Mouse-eared cress

**SCIENTIFIC NAME:** *Arabidopsis thaliana*

**COUNTRY OF ORIGIN:** A common weed in Europe, North America, and parts of Asia

**INTERESTING FACTS:** This plant may seem small and not very pretty, but it is very important. It is the first plant to have all of its genes discovered. Scientists chose it because it is so easy to grow and makes lots of seeds.

**Scientists now have a list of every gene in this plant's nucleus.**

## If scientists could put a gene from fireflies into a plant, the plant would glow in the dark.

In the future, scientists will be able to take specific genes from one plant and put them directly into a different plant. They may be able to take a gene from a plant that grows in a cold climate and put it into a plant that grows in a warm climate. Then plants that usually grow only in warm places could grow just about anywhere.

For example, a gene that allowed a plant to survive cold weather could be put into an orange tree, which needs a warm climate to produce oranges. Thanks to plant genetics, oranges, which now grow in California and Florida, could grow in northern Canada!

Can you imagine growing oranges in Canada?

**THINK IT OVER!**

**Brainstorm another new plant scientists might create by taking a gene from another living thing. What plant would you start with? What gene would you add? Write a description of your new plant and what it would do.**

Scientists may also be able to create new plants that have never been seen before. For example, they might take the gene for blue flowers out of one plant and put it into the cells of a rose plant.

Scientists may even be able to put genes from animals into plants. If they could put a certain gene from fireflies into a plant, the plant would glow in the dark. Imagine how your front yard or the park where you play would look at night if all the trees glowed in the dark!

Can you imagine giving someone a bouquet of blue roses?

**Some people think that scientists should not be mixing the genes of different plants until they understand more about plant genetics.**

Scientists may also be able to make food taste better or be healthier. For example, in some countries, people eat a lot of rice. Rice is good to eat, but it does not contain vitamin A, which people need for healthy eyes. Carrots contain lots of vitamin A, so scientists are taking genes from carrots and putting them into rice. The new kind of rice they are growing is called golden rice. Golden rice will feed millions of people who do not get enough vitamin A in their food.

"Golden rice" was produced by putting genes from carrots into white rice.

However, some people think that scientists should not be mixing the genes of different plants until they understand more about plant genetics.

For example, scientists are working on creating plants that contain a chemical that kills insects. These plants would not have to be sprayed with poisonous chemicals to keep insects from eating them. But some people are afraid that these plants may kill all insects, which would upset the balance of nature. Or the fruits and vegetables of these plants might be dangerous for humans to eat.

People also worry that we might accidentally create super plants that could turn into weeds. These super weeds could invade and take over farmers' fields or people's gardens.

**Man-eating plants were the stars of the Broadway show *Little Shop of Horrors.***

Scientists who study plant genetics will continue to gather new information. Already they have helped to explain the birth of agriculture. They have shown us how plants grow and why some plants are so different from others. They have even experimented to create tastier, more nutritious foods. The next time you look at a plant, think about how plant genetics has helped to reveal the many mysteries locked inside a plant's cells.

**QUESTION:**

**How do you think plant genetics will be used in the future?**

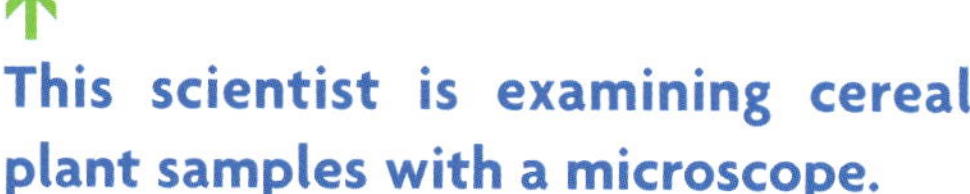

This scientist is examining cereal plant samples with a microscope.

# Glossary

**agriculture** growing plants for food; another word for farming

**ancestor** a plant from long ago that is related to a present-day plant

**anther** part of a flower where *pollen* is made

**cells** microscopic, square-shaped boxes that make up all living things

**crossing** taking *pollen* from one plant and putting it on another plant to make seeds

**dominant** stronger; opposite of *recessive*

**DNA** chemical that *genes* are made of

**genes** special sets of instructions inside a *cell* that tell a plant or animal how to grow

**genetics** the science of *genes*

**hybrid** a new plant made by *crossing* one kind of plant with a different kind of plant

**molecules** small, nearly invisible particles that make up chemicals; for example, molecules of salt or molecules of *DNA*

**nucleus** the control center of a plant *cell* with *genes* inside

**ovary chamber** part of a flower where the seeds are made

**pollen** dust-like, yellow *cells* made in a flower's *anther*

**recessive** weaker; opposite of *dominant*

**stigma** sticky, green part of a flower that traps *pollen*

# Index